ALTERNATOR BOOKS™

WEIRD PLACES

Jackie Golusky

Lerner Publications ◆ Minneapolis

Copyright © 2024 by Lerner Publishing Group, Inc.

All rights reserved. International copyright secured. No part of this book may be reproduced, stored in a retrieval system, or transmitted in any form or by any means—electronic, mechanical, photocopying, recording, or otherwise—without the prior written permission of Lerner Publishing Group, Inc., except for the inclusion of brief quotations in an acknowledged review.

Lerner Publications Company
An imprint of Lerner Publishing Group, Inc.
241 First Avenue North
Minneapolis, MN 55401 USA

For reading levels and more information, look up this title at www.lernerbooks.com.

Main body text set in Aptifer Sans LT Pro.
Typeface provided by Linotype AG.

Editor: Brianna Kaiser **Designer:** Mary Ross

Library of Congress Cataloging-in-Publication Data

Names: Golusky, Jackie, 1996– author.
Title: Weird places / Jackie Golusky.
Description: Minneapolis, MN : Lerner Publications , [2023] | Series: Wonderfully weird
   (Alternator Books) | Includes bibliographical references and index. | Audience: Ages 8–12
   years | Audience: Grades 4–6 | Summary: "All around the world, there are some strange and
   unique places. From human-made landscapes to natural wonders, readers will discover
   some of the odd places people have explored"— Provided by publisher.
Identifiers: LCCN 2022033593 (print) | LCCN 2022033594 (ebook) | ISBN 9781728490717 (Library
   Binding) | ISBN 9798765601839 (eBook)
Subjects: LCSH: Geography—Miscellanea—Juvenile literature.
Classification: LCC G133 .G65 2023 (print) | LCC G133 (ebook) | DDC 909—dc23/eng20221020

LC record available at https://lccn.loc.gov/2022033593
LC ebook record available at https://lccn.loc.gov/2022033594

Manufactured in the United States of America
1-53002-51020-11/30/2022

TABLE OF CONTENTS

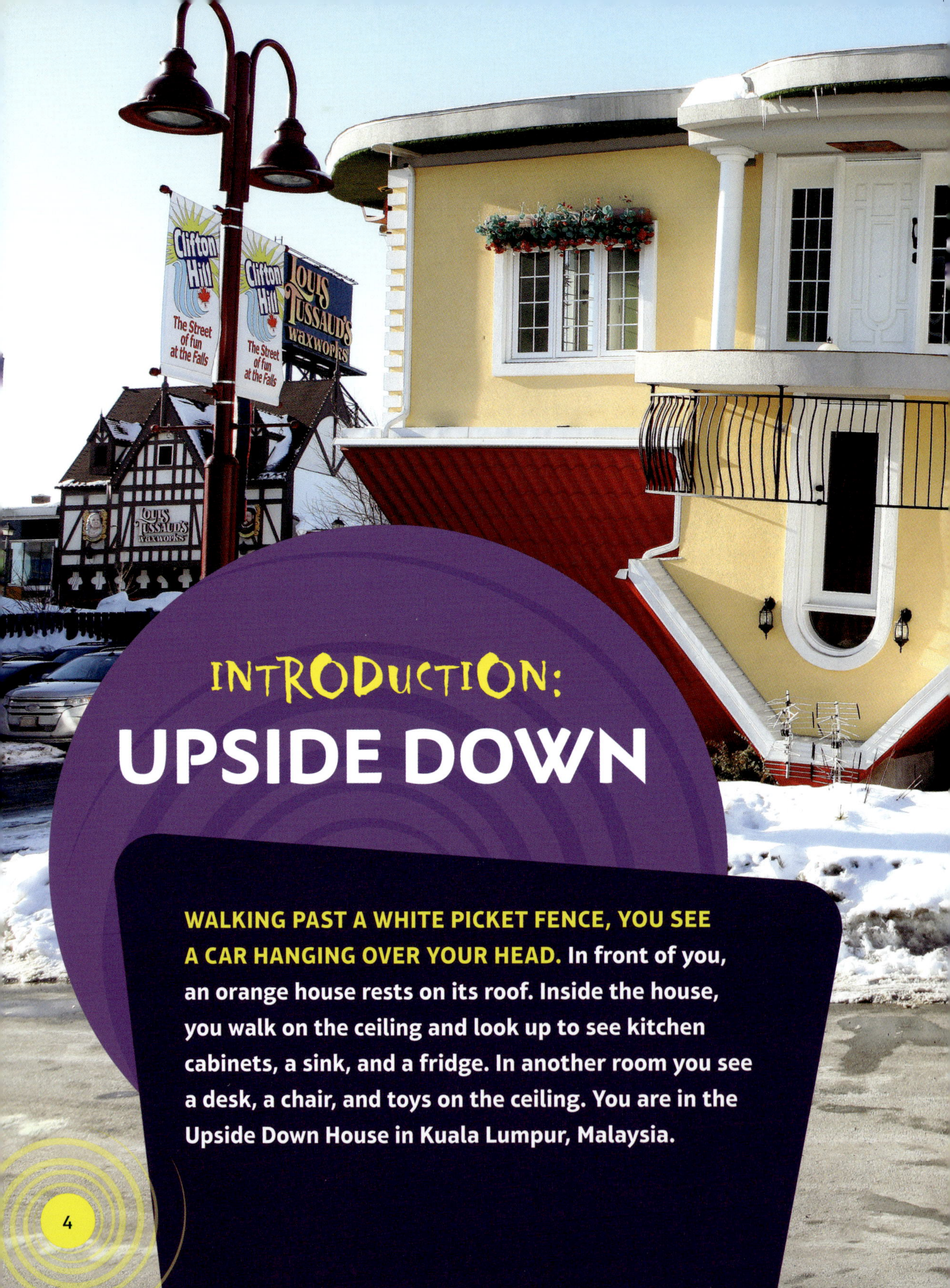

INTRODUCTION:
UPSIDE DOWN

WALKING PAST A WHITE PICKET FENCE, YOU SEE A CAR HANGING OVER YOUR HEAD. In front of you, an orange house rests on its roof. Inside the house, you walk on the ceiling and look up to see kitchen cabinets, a sink, and a fridge. In another room you see a desk, a chair, and toys on the ceiling. You are in the Upside Down House in Kuala Lumpur, Malaysia.

4

This home is one of the handful of the world's buildings that have been built upside down. You can find other upside-down houses in many countries, including Thailand, Germany, Russia, and Canada. An upside-down White House is even in Wisconsin Dells, Wisconsin. These buildings were carefully constructed and decorated to look as though they are upside down. People walk through them and enjoy the strangeness of them. In addition to the Upside Down House, you can find more weird places around the globe.

CHAPTER 1:

WEIRDLY COLORFUL

FROM PINK WATERS TO ROWS OF RED PLANTS, THESE SITES STUN WITH THEIR STRANGE COLORS:

LAKE HILLIER

Australia's Lake Hillier wows people with its pink hues. If you scooped its water into a clear glass cup, the water would still appear pink. That's not true for the rest of the world's pink lakes.

Lake Hillier stays pink all year. Scientists don't know why. Some scientists think it's because of a certain algae in the lake. These plants with no leaves or stems grow near water. The algae could be making a pigment that causes the pink coloring. Other scientists think the lake's bacteria and salt are reacting to each other, creating the lake's color.

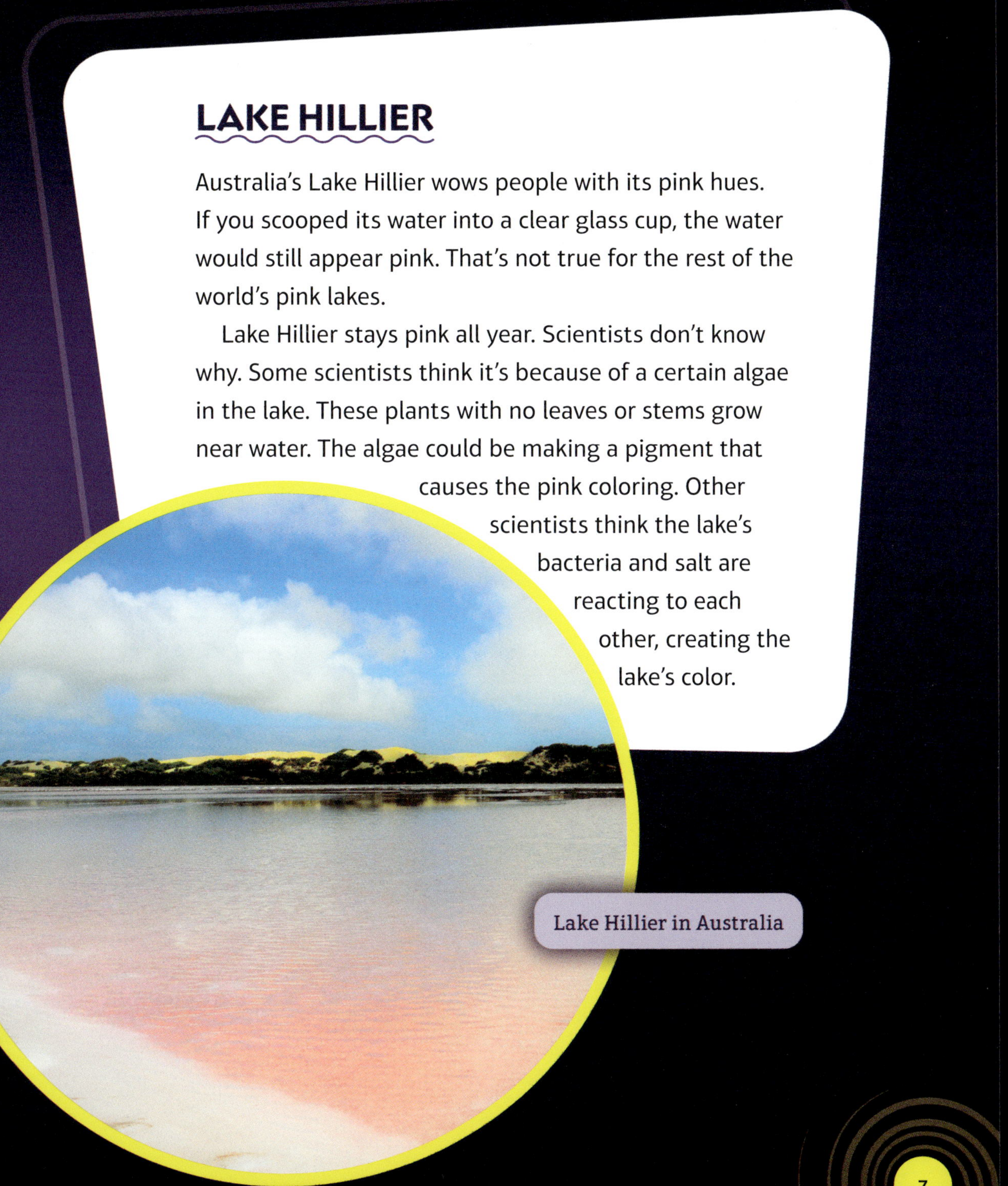

Lake Hillier in Australia

OTHER PINK LAKES

Lake Hillier isn't the world's only pink lake. Australia is also home to Quairading Pink Lake. A road divides the lake in two. The different sides of the lake often are different shades of pink.

Senegal has its own pink lake, Lake Retba. Algae turned Retba's waters pink. It is also one of the world's saltiest lakes.

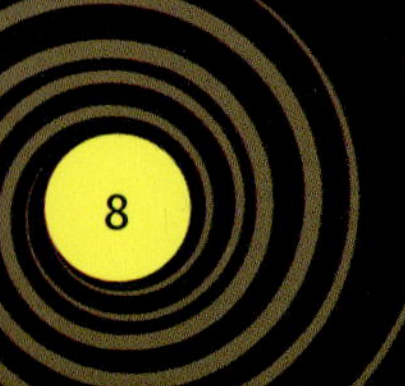

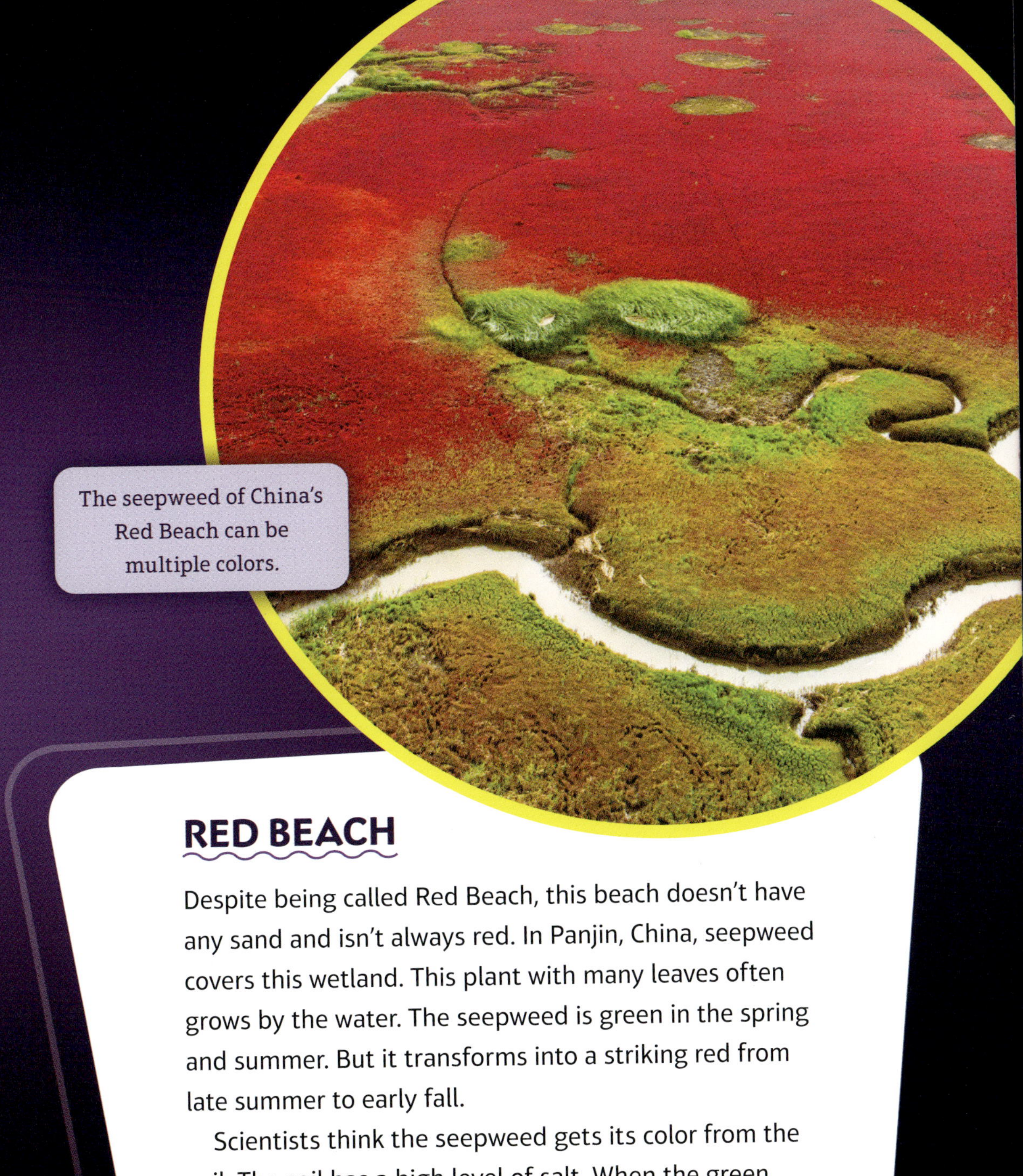

The seepweed of China's Red Beach can be multiple colors.

RED BEACH

Despite being called Red Beach, this beach doesn't have any sand and isn't always red. In Panjin, China, seepweed covers this wetland. This plant with many leaves often grows by the water. The seepweed is green in the spring and summer. But it transforms into a striking red from late summer to early fall.

Scientists think the seepweed gets its color from the soil. The soil has a high level of salt. When the green seepweed absorbs the salt from the soil, it turns from green to red before dying in the fall.

BLOOD FALLS

Bright red water flows down Blood Falls in Antarctica. The water looks like blood. For a long time, scientists didn't know why the falls were red. They discovered it is due to the high level of iron in the falls' salt water. This metal appears in many things including blood. When the iron in the salt water is exposed to oxygen, the iron turns red, making Blood Falls appear red too.

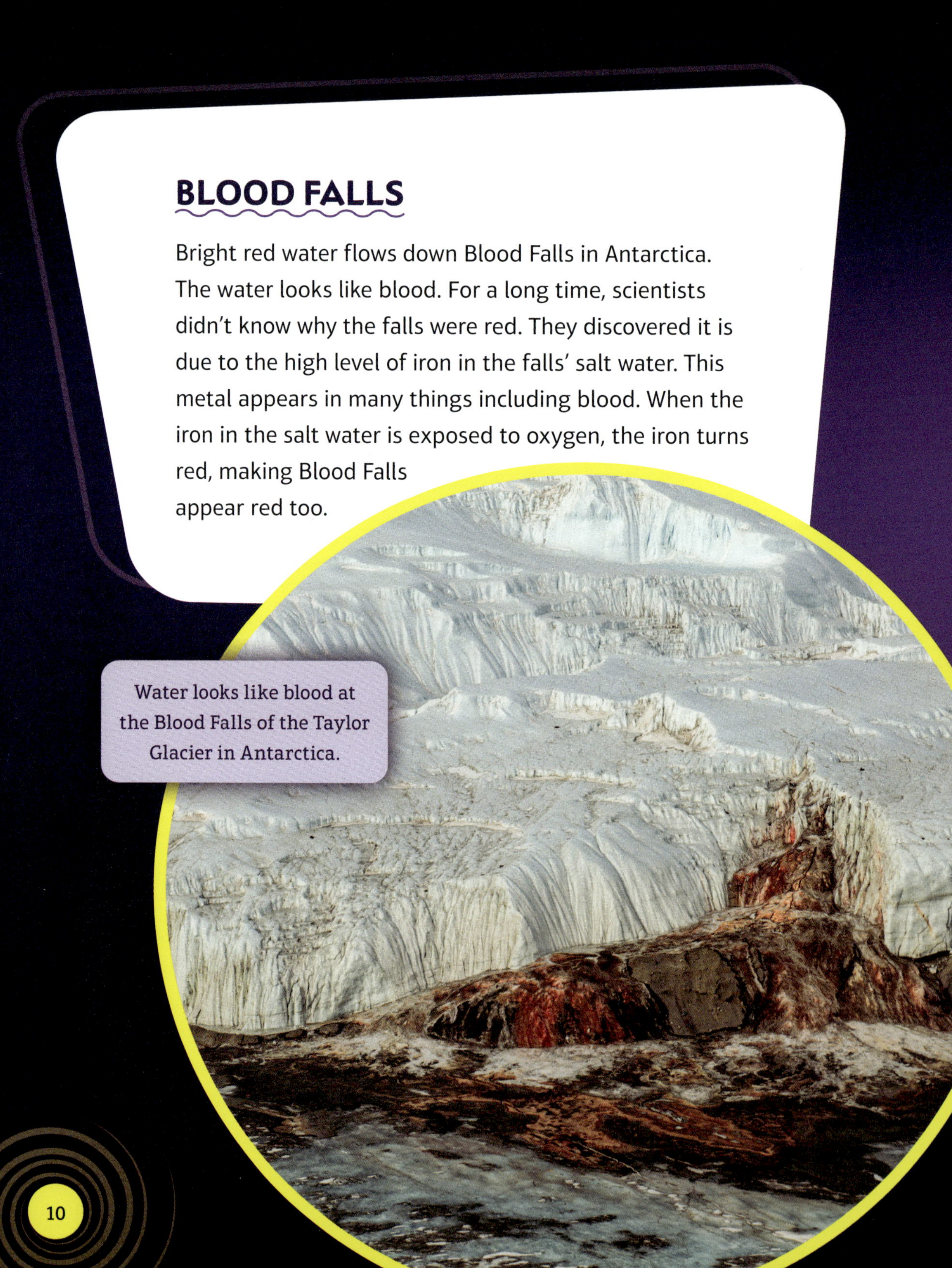

Water looks like blood at the Blood Falls of the Taylor Glacier in Antarctica.

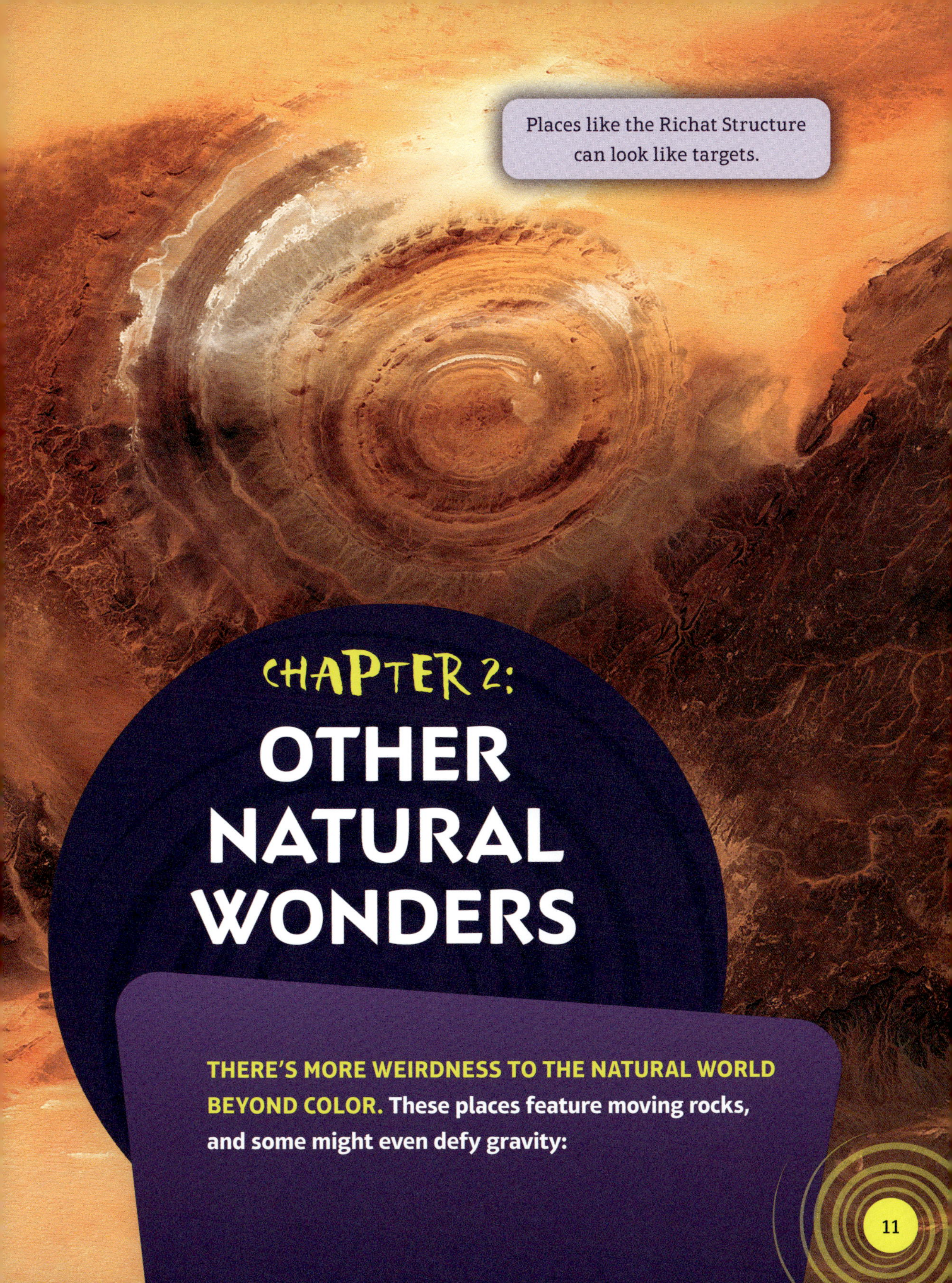

CHAPTER 2:
OTHER NATURAL WONDERS

THERE'S MORE WEIRDNESS TO THE NATURAL WORLD BEYOND COLOR. These places feature moving rocks, and some might even defy gravity:

DEATH VALLEY

Death Valley is a dangerously hot desert, yet it seems to bring rocks to life. On the border between California and Nevada, Death Valley is home to the sailing stones. These rocks have moved over the years, but no one has seen them move.

We know these rocks move because they leave trails behind them. People have also seen the rocks appear in different locations. The stones weigh from several ounces to hundreds of pounds.

Some scientists think ice, water, and wind move the rocks. They believe water sometimes freezes overnight and melts the next day. Then the wind breaks up the thin pieces of ice, which push the rocks.

MAGNETIC HILL

The Magnetic Hill in Ladakh, India, might defy gravity. If someone parks a car on the hill, the car will go up the hill 12 miles (20 km) per hour. People wonder why Magnetic Hill pulls things upward. That's why some call it Mystery Hill.

Some scientists think that the hill has a strong magnetic force. Then the hill's force would pull things such as a car toward it. Other people think it is an optical illusion—a trick of the eye making something appear to be something else. They believe that the car is actually moving downhill and the hill just makes it appear as if the car is climbing uphill.

No-Fly Zone

The Indian Air Force doesn't fly over the Magnetic Hill. Pilots worry its magnetic force will disrupt their equipment.

The Magnetic Hill in India appears to pull things upward.

RICHAT STRUCTURE

The Richat Structure in Mauritania, Africa, is easier to see in space than it is on the ground. It is 31 miles (50 km) across with huge circles that make it look like a target. Astronauts wonder where these perfect circles came from.

One theory is that it is a landing site for aliens. Another theory is that an asteroid hit Earth and the impact of the asteroid created the circles. Scientists believe that the site is caused by erosion. Erosion occurs where something is worn away by nature. These scientists think that wind and water make the weaker rocks erode. Then the stronger rocks are left, creating the circle ridges.

ETERNAL FLAME

A flame burns behind a waterfall in Chestnut Ridge Park in Orchard Park, New York. Even when the falls freeze, the fire still burns. The flame has gone out several times, though it has always been relit by people or relites on its own.

Researchers know that the flame burns on top of gas in the rock. The gas gives the fire the fuel it needs to burn. But what they don't know is how the gas got there. Some people think the gas comes from a really hot rock, but the rocks around the Eternal Flame aren't that warm. Researchers are trying to learn more about this mysterious site.

BURNING UP

There are more eternal flames in the world, but scientists know why they keep burning. Some are caused by extremely hot shale, a type of rock that often has gas in it. In Pennsylvania, a fire burns continuously in Cook Forest State Park. But the gas that keeps the fire burning is from an old gas well.

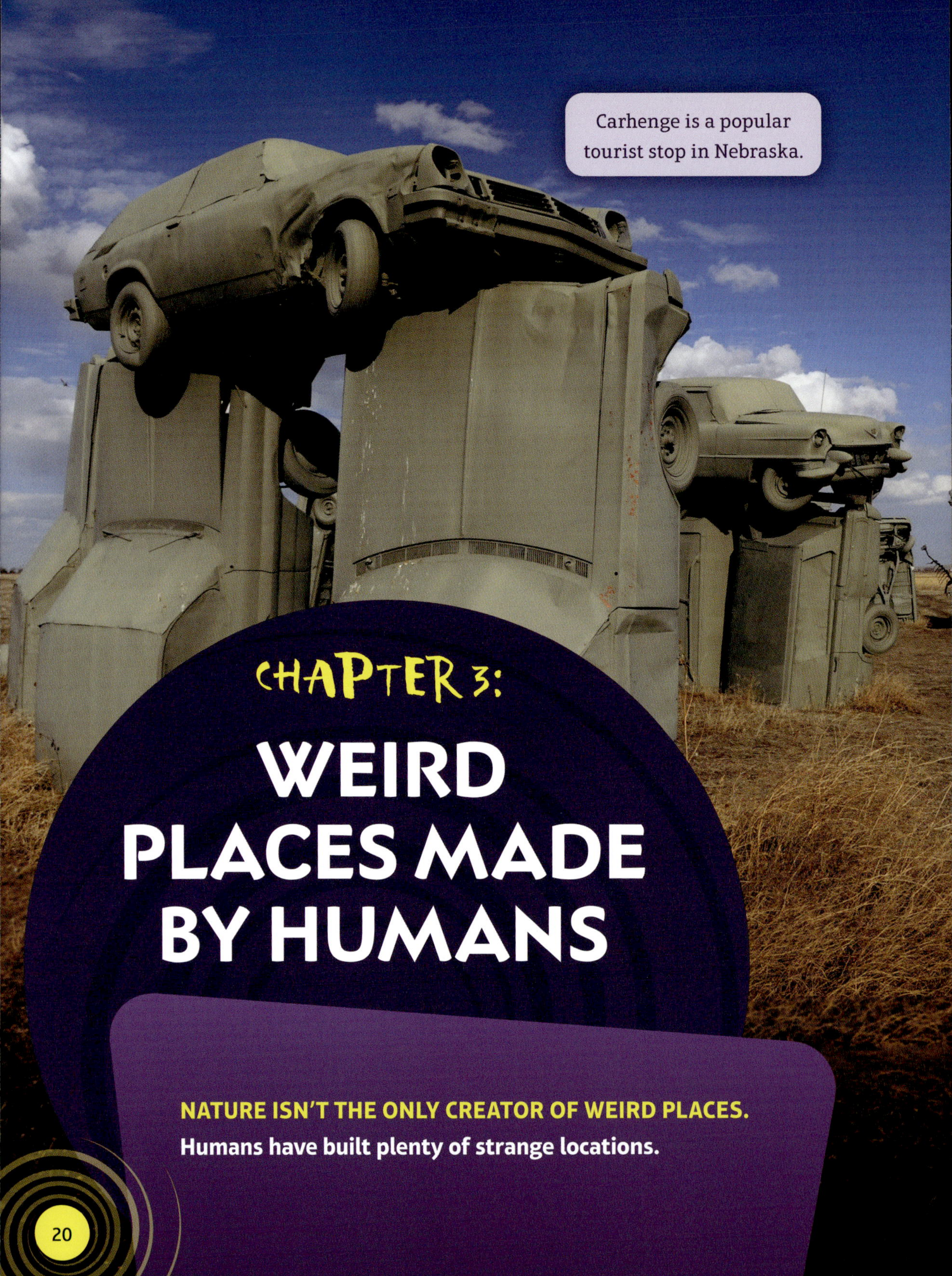

CHAPTER 3:
WEIRD PLACES MADE BY HUMANS

NATURE ISN'T THE ONLY CREATOR OF WEIRD PLACES.
Humans have built plenty of strange locations.

PAPER HOUSE

In Rockport, Massachusetts, the Paper House is made from just that: paper. The walls and ceilings have 215 layers of paper that are stuck together with flour, water, and apple peels. Engineer Elis Fritiof Stenman built this house in 1924 because he was curious whether a whole house could be built out of paper.

A piano in the Paper House is covered with newspaper. It is the only furniture in the house not made entirely of newspaper.

The Paper House is in Rockport, Massachusetts.

Stenman loved reading the newspaper and used some of his newspapers for his house. Many people donated newspapers too. The house was made from about one hundred thousand sheets of newspaper. Even the furniture is made from them. Stenman's wife, Esther, created curtains for the house from magazine covers. This means you can read the house, its furniture, and its curtains.

A Castle of Corn

South Dakota's Corn Palace is decorated with corn, grains, and grasses. It's redecorated with twelve colors of corn every year. All the colors, including red and blue, are natural corn colors.

TOILET AMUSEMENT PARK

What is one invention that has changed your life? Most people wouldn't say the toilet, but it has greatly changed people's lives. There's a theme park called the Restroom Cultural Park devoted to this invention in South Korea.

Sim Jae-duck created the park. He improved public toilets when he was the mayor of Suwon, South Korea. He also built and lived in the first house in the world that was shaped like a toilet. After he died, his house was made into a museum. People stroll through the toilet-shaped museum to learn about the history of the toilet, how it came to be, and the many versions of it, including chamber pots.

Sim Jae-duck's toilet-shaped house

CHEWING GUM WALL

Ever need to get rid of your gum but don't see any trash cans? In Seattle, Washington, a wall full of gum started in the 1990s when people stuck their gum on the wall instead of putting it in a trash can. More and more pieces were stuck to the wall. Now the 15 by 50 foot (4.6 by 15 m) wall is covered in globs of gum. If you visit, you can add your own sticky glob to the wall.

FLAVOR GRAVEYARD

Many people are upset when they can't find their favorite ice cream flavors. Sometimes companies such as Ben & Jerry's stop making certain flavors. Ben & Jerry's has dedicated a cemetery to the flavors that the ice cream company has stopped making. The cemetery, called the Flavor Graveyard, is part of the company's factory tour in Vermont. Gravestones bear the names of former flavors such as Peanut Butter and Jelly, Sugar Plum, and Chocolate Comfort.

The Flavor Graveyard remembers old flavors of Ben & Jerry's ice cream.

CADILLAC RANCH

Off Interstate 40 near Amarillo, Texas, ten Cadillacs stand with their noses buried in the dirt. Graffiti covers these cars. Artists made the site in 1974. Visitors come from all around the world to see this strange display. Many people spray-paint the cars with spray paint that's already there or bring their own. You can visit this site and add your own mark to it.

Carhenge

Near Alliance, Nebraska, a set of cars are partially buried and stand on top of one another to resemble Stonehenge. Stonehenge is a historic structure in England.

MARKEL BUILDING

Often referred to as one of the ugliest buildings in the world, the Markel Building makes its presence known. The Markel Building is in Richmond, Virginia. Its three sections of metal look like crumpled aluminum foil. The creator of this building was inspired by a baked potato in foil. The building's metal gets its texture from people beating sledgehammers into it. Over the years, the metal has been patched with duct tape.

The Markel Building in Virginia

MORE WEIRDNESS

There are plenty of weird places around the world. This book highlights just a few. There's also Thor's Well in Oregon where water pours from the ocean into a sinkhole. Aoshima Island is one of several cat islands in Japan. For every person on the island, there are six cats. What are some weird places you've visited or want to visit?

Japan has multiple cat islands in addition to Aoshima.

GLOSSARY

decorate: to try to make something nicer by putting something on it

dedicate: to do something to remember or honor a person, event, or thing

engineer: a person who designs and builds things

gas: something that is like air and has no fixed shape

gravity: a natural force that causes things to move toward each other

hue: a color or a shade of color

magnetic: a stone or piece of iron or steel that attracts bits of iron or steel

pigment: something that gives color to animals and plants

texture: the way that something feels when you touch it

LEARN MORE

Britannica Kids: Death Valley
https://kids.britannica.com/kids/article/Death-Valley/353041

Burgan, Michael. *Rocks and Minerals*. Washington, DC: National Geographic Kids, 2022.

The Corn Palace
https://cornpalace.com/27/About-Us

Doeden, Matt. *Travel to Australia*. Minneapolis: Lerner Publications, 2022.

Flannery, Tim. *Weird, Wild, Amazing! Desert: Exploring the World's Incredible Drylands*. New York: Norton Young Readers, 2022.

Kaiser, Brianna. *Weird Plants*. Minneapolis: Lerner Publications, 2024.

Kiddle: Suaeda Maritima Facts for Kids
https://kids.kiddle.co/Suaeda_maritima

The Paper House
https://www.paperhouserockport.com/inside.html

INDEX

PHOTO ACKNOWLEDGMENTS

Image credit: Raymond Boyd/Getty Images, pp. 4, 5; DigitalGlobe/Getty Images, p. 6; Chloe Vid/Shutterstock, p. 7; HelloRF Zcool/Shutterstock, p. 9; Cavan Images/Alamy Stock Photo, p. 10; Trismegist san/Shutterstock, p. 11; Anatoliy Lukich/Shutterstock, pp. 12, 13; Kieu images/Shutterstock, pp. 14, 15; GLF Media/Shutterstock, pp. 16, 17; Jason Ondreicka/Alamy Stock Photo, p. 18; Paul Harris/Getty Images, p. 20; John Blanding/The Boston Globe/Getty Images, p. 21; Norman Barrett/Alamy Stock Photo, p. 22; JUNG YEON-JE/AFP/Getty Images, p. 23; Buiobuione/Wikipedia, p. 24; Peter Conner/Alamy Stock Photo, p. 25; Josh Brasted/Getty Images, p. 26; Jim/Wikipedia, p. 27; The Asahi Shimbun/Getty Images, pp. 28, 29.

Cover: Josh Brasted/Getty Images.